Killer Comet

What the Carolina Bays tell us

Antonio Zamora

ANTONIO ZAMORA

This work tries to reconstruct what happened at the end of the last ice age when a land bridge joined Asia and America allowing humans to migrate. Did the humans kill the megafauna that populated the continent, or was their disappearance the result of a natural catastrophe?

ISBN: 978-0-9836523-7-3
Paperback Edition 3

CONTENTS

INTRODUCTION

During the Last Glacial Maximum, about 22,000 years ago, Canada and the northern United States were covered with ice sheets up to two kilometers thick. The Laurentide ice sheet extended approximately from the Rocky Mountains to the Atlantic Ocean, and the Cordilleran ice sheet covered the westward side of the mountain range all the way to the Pacific Ocean. With so much of the world's water stored as ice over the Earth's crust, the sea level was lower than today, and a land bridge called Beringia joined Siberia and Alaska. Toward the end of the ice age, 15,000 years ago, a corridor opened across Canada between the ice sheets that allowed animals and humans to migrate from Siberia to America. Many anthropologists have suggested that some the first human settlers of America came across this corridor and this is why Native Americans have many Asian genetic markers (Dulik, 2012).

When humans arrived during the ice age, North America already was inhabited by large animals or megafauna that included giant sloths, short-faced bears, American lions, saber-toothed tigers, camels, mammoths and mastodons. All the big animals became extinct around the time when humans populated America. The expansion of humans into previously unpopulated regions, such as Australia, has been generally accompanied by extinctions of other species (Burney, 2005). Some paleontologists have proposed that humans, acting as superpredators killed key species and modified the environment by the use of fire, thereby causing the demise of the megafauna.

As evidence of this, they point out the distinctive fluted stone arrowheads crafted by the Clovis people that have been found at many archeological sites containing bones of big game scratched by butchering, including mastodons, mammoths, camels, horses and giant tortoises.

Not all scientists attribute the extinction of the megafauna to overhunting and loss of habitat caused by human settlement in North America. Other possible causes of the extinctions could have been the spread of infectious diseases brought by the new immigrants and large changes in climate patterns. There is not much evidence to support a wide extinction due to infectious diseases that could have affected multiple species with different immune systems. The extinction of the megafauna coincided with the onset of the Younger Dryas cold event, which was a period of sudden cold climatic conditions starting approximately 12,900 years ago and lasting 1,300 years. It is not clear why species that had endured the temperatures of the Last Glacial Maximum would not have adapted to the new conditions.

Paleontologists discuss primarily overhunting and climate change as the main reasons for the megafauna extinction, but they favor climate change as the dominant driver of most biotic phenomena. However, some scientists are considering that an impact from an asteroid or comet, smaller than the one that killed the dinosaurs, could have killed the North American megafauna and the Clovis people.

Richard B. Firestone of the Lawrence Berkeley National Laboratory and some of his collaborators proposed in 2007 the hypothesis that the explosion of an extraterrestrial object over the ice sheets of Canada around 12,900 years ago caused the megafaunal extinction and triggered the Younger Dryas cold event. This hypothesis, called the Younger Dryas Impact Hypothesis, has been the subject of contentious debate because the site of the putative explosion has not been found, and spikes in the concentration of rare elements from a meteorite, such as iridium, have not been found in the layer of soil associated with the extinction, as was the case with the impact that killed the dinosaurs.

Proponents of the impact hypothesis argue that an airburst or an impact on the thick glaciers would have shielded the Earth's surface to prevent the formation of a crater and that erosion by the water from the melting glaciers could have easily concealed the location of the impact zone. Firestone's evidence for an extraterrestrial impact consists of microscopic hexagonal diamonds that can only be created at great pressures and microscopic spherules of organic and inorganic matter that may have originated from the extraterrestrial collision and forest fires ignited by the heat of the impact. The elliptical Carolina Bays, which also have been suggested as being the result of the proposed extraterrestrial impact, do not contain meteorite fragments or any of the characteristics associated with extraterrestrial impacts, such as quartz crystals deformed by the high pressures of an impact, although microspherules attributed to the impact have been found within them. Opponents of the Younger Dryas Impact Hypothesis, which include many extraterrestrial impact experts, have not accepted the nanodiamonds and microspherules as evidence of a cosmic impact. In addition, soil samples from the terrain in which the Carolina Bays are found have produced a wide range of ages, and this has caused geologists to discard an impact hypothesis and to consider only that the Carolina Bays were formed over thousands of years by water or wind geomorphic processes.

The Younger Dryas Impact Hypothesis is appealing because it provides a neat explanation of several events that happened approximately at the same time: 1) the extinction of the megafauna, 2) the disappearance of the Clovis culture from the fossil record, and 3) the onset of the Younger Dryas cold event. Unfortunately, the evidence submitted thus far in support of this hypothesis has not been sufficient to satisfy the critics who demand unambiguous scientific proof.

The impact origin of the Carolina Bays has been a contentious issue since their discovery in the 1930s. Given this prolonged controversy, any book trying to associate impacts to the megafauna extinction and the Younger Dryas event is likely to be received with great skepticism. Nevertheless, the premise of this book is that an extraterrestrial object, such as an asteroid or a comet, was responsible for the extinction of the megafauna.

The introduction of new human populations to North America and the elimination of large mammalian herbivores and destruction of their habitat are often called the "Pleistocene overkill" hypothesis. Paleontologists do not think that the coincidence of human migration and megafauna extinction presents a convincing case for attributing the large extinction event to human predation. The next chapters provide a brief background of what is known about the megafauna and the people who inhabited North America during the ice age.

This book introduces the Glacier Ice Impact Hypothesis, which describes a sequence of four mechanisms by which the Carolina Bays could have formed after an extraterrestrial impact on the ice sheet that covered North America. The mechanisms are ordinary geological processes, but they have not been considered in the particular sequence presented here. Photographs of experimental tests demonstrate that rigorous scientific principles underlie the new hypothesis. Information is presented about the characteristics of extraterrestrial impacts and about the Carolina Bays.

The calculations derived from the Glacier Ice Impact Hypothesis and the geographic distributions of the Carolina Bays provide information about the magnitude of the extraterrestrial impact and its effect on the glacier ice sheet. Understanding how the Carolina Bays formed helps to explain how the material ejected by an extraterrestrial impact could have led to the extinction of the megafauna and acted as a trigger for the Younger Dryas cooling event.

***

THE LAST ICE AGE

The most recent glacial period occurred toward the end of the Pleistocene Epoch, approximately from 110,000 to 12,000 years ago. This "ice age" was the latest glaciation event of several glacier advances and retreats during the previous two million years. The Last Glacial Maximum, when ice reached its maximum coverage of land, was approximately 22,000 years ago. Different patterns of glacier advance and retreat occurred in different continents. The Weichselian glaciation froze the Baltic Sea and covered northern Europe with ice, including Norway, Sweden, Finland and the northern parts of England, Germany and Poland. The Wisconsin glaciation in North America was a southward extension of the Laurentide ice sheet that covered most of Canada and advanced toward the Upper Midwest and New England. At its maximum, the glacier ice covered the Great Lakes, all of Michigan and most of North Dakota, Minnesota and Wisconsin. The glacier ice also covered the northern parts of Indiana and Ohio. All the coastal states north of Pennsylvania were completely covered with a layer of ice up to two kilometers thick. New York's Central Park still has some rock outcrops with the grooves left by glaciers that scraped the surface as they made their way to the Atlantic Ocean.

How do we know what happened so long ago? Geologists examine the layers of the earth to reconstruct the events. Deep layers are deposited before the top layers, and the dates of the layers can be determined from the ratios of chemical isotopes

9

for which the rate of decay is known. Clues can be obtained from the physical processes that change the land, such as erosion or the freezing and thawing of water. When snow falls on the ground and becomes deeply packed to form glaciers, the stones from the landscape become imbedded in the ice and they can grind the surface when the glacier moves. After the ice finally melts, the stones that are imbedded in the glacier drop to the ground where they may remain in piles or are sometimes integrated in mud that transforms into sedimentary rock. Geologists look for the grooves carved by the glaciers and for piles of glacial rock debris to determine the extent of past ice coverage. The chemical composition of the rocks can be used to determine their origin and establish the direction in which the glaciers moved. This is how we know that glaciers once covered all of Michigan and Minnesota, as well as the northern portions of Ohio, Indiana and Illinois. Every time that the ice retreated due to warmer temperatures, the glaciers melted and left a trail of stones.

Everything that happened on the Earth before the invention of writing about 5000 years ago can only be deduced through the disciplines of geology, archaeology and paleontology. The dating of fossils, bones, artifacts and soil samples can be used to try to produce a coherent story of the past, but disturbance of the soil and conflicting evidence can make it very difficult to find out what really happened.

Modern humans, *Homo sapiens,* evolved as a species approximately 160,000 years ago. They were still in Africa at the start of the last ice age using Paleolithic stone tools comparable to those made by

Neanderthals in Europe and by *Homo erectus* in Asia. Toward the end of the ice age, *Homo sapiens* had colonized the whole planet and had become the only surviving human lineage. By then, they were using finely crafted flaked Mesolithic stone tools.

The disappearance of the large animals that inhabited North America at about the time when modern humans populated the continent has received much attention, but finding a major cause for the demise of the animals has been very elusive. Some attempts have been made to use the oral histories and mythologies of Native Americans to find out what happened thousands of years ago, but this approach has not provided information that can be verified objectively. Analysis of the DNA from remains of ancient fauna has also been used to determine the migration patterns of the animals that inhabited North America during the ice age and of the humans who hunted them. Even with all these resources, paleontologists do not agree whether the megafauna were annihilated by overhunting, by changes in the climate, by the spread of infection, or, as more recently suggested, by an extraterrestrial impact substantially smaller than the one that killed the dinosaurs throughout the world at the end of the Cretaceous Period.

Understanding the chronology of major events during the last ice age is necessary to determine how America was populated and how humans affected their environment. It has been difficult to come to a consensus about the extinction of the megafauna because much of the information is spotty and there is no way of selecting between competing theories

11

when there is insufficient information to choose one over the other.

During the Last Glacial Maximum, modern humans populated Africa, Europe and Asia. They survived the cold weather by using controlled fires and by making clothing from the skins of the animals that they hunted. These cold-adapted humans would soon colonize America. Siberia and Alaska were connected by a land bridge called Beringia that was 1000 kilometers wide. It is believed that a small human population from Siberia settled in eastern Beringia, which is now Alaska, before expanding toward South America sometime after 16,500 years ago. Some scientists think that humans could also have come along the Pacific coast in boats like those used by the Inuits.

The Cordilleran ice sheet on the Pacific coast separated from the Laurentide ice sheet around 15,000 years ago. The corridor that opened in Canada between the ice sheets made it possible for animals and humans to migrate southward to populate America. Some animals that were native to America, such as camels and horses, crossed the Beringia land bridge toward Asia.

When the climate became warmer, the large North American glaciers started melting and increased the sea level. Beringia was inundated by the rising sea and any human settlements that may have been along the coastlines were covered by water up to 130 meters deep, making them inaccessible to researchers.

Large animals had inhabited all of North America for thousands of years, and suddenly, the large animals and the newly arrived Clovis people who hunted them disappeared around 12,900 years ago. The climate got colder for approximately 1300 years in what is known as the Younger Dryas cold event. The end of the Pleistocene Epoch, 11,400 years ago, marks the end of the ice age and the start of the Holocene Epoch, which is our current epoch.

Today, paleontologists are trying to answer questions related to the late Pleistocene megafauna extinction. Spear points have been found at sites where camels, mammoths and the large *Bison antiquus* were butchered. Is it possible that the stone-age people of North America hunted all the animals to extinction? Are the humans that populated North America after the ice age the descendants of the Clovis people? Or, did the Clovis people also disappear along with the megafauna and were replaced by a different line of humans?

The following are some of the large animals, both prey and predators, that disappeared at the end of the ice age (Grayson, 2002).

Bison antiquus was a direct ancestor of the living American bison although it was much larger. It was the most common herbivore in North America from 18,000 to about 10,000 years ago.

A large armored herbivorous armadillo, *Doedicurus clavicaudatus,* inhabited woodlands and grasslands. This animal weighed about two metric tons and was 1.5 meters high.

The North American camel, genus *Camelops*, roamed North America until its disappearance about 10,000 years ago.

The woolly mammoth (*Mammuthus primigenius*) coexisted with early humans in North America. It was hunted by humans, and its bones and tusks were used for tools and as posts for dwellings. Its disappearance in North America at the end of the Pleistocene 10,000 years ago is thought to have happened as a consequence of hunting and loss of habitat because mammoths were still living in Siberia until 4,000 years ago.

Large cats were some of the most common predators during the Pleistocene epoch. The American cheetah (*Miracinonyx*) was similar to the modern cheetah and it disappeared at the end of the ice age.

The American lion (*Panthera leo atrox*) was endemic to North America and northwestern South America for over 300,000 years until its disappearance 11,000 years ago. The American lion was about 25% larger than the modern African lion.

Smilodon, a robust cat equipped with large canine teeth, was a specialized hunter of bison and camel. Many fossils of *Smilodon* have been recovered from the La Brea Tar Pits in Southern California. This big cat also disappeared from North America at the end of the ice age.

These are just some examples of the multitude of large animals that became extinct during the transition from the Pleistocene to the Holocene epoch. In North America, around 45 of 61 genera of large mammals weighing more than 45 kilograms (100 pounds) became extinct. South America and Australia also had extinctions of large animals during the Pleistocene, but there is uncertainty about the timing of the events and whether these extinctions had the same causes as the North American

14

extinction. Paleontologists have suggested that the combination of rapid climate change and overhunting by humans contributed directly or indirectly to the demise of the megafauna (Prescott, 2012). Something that casts doubt on the human overkill hypothesis is that comparable extinctions did not occur in Africa and Southeast Asia where large fauna evolved alongside humans.

Smilodon, the saber-toothed cat

***

HUMAN MIGRATION TO AMERICA

Early in the 20th century, several sites were found containing distinctive stone spear points and arrowheads with fluted bases that showed meticulous craftsmanship. The tools dated from about 13,500 to 13,000 years ago. The makers of these stone tools were named the Clovis people because the tools had been found at an excavation site near Clovis, New Mexico. The widespread distribution of Clovis-style artifacts meant that the people who had created these tools had successfully settled a large portion of the United States with their Mesolithic technology.

Clovis projectile point made by pressure flaking on both sides. The base is fluted for mounting on a shaft. Image courtesy of the Virginia Dept. of Historic Resources.

For many years it was thought that the Clovis people had been the first human inhabitants of America and that they had arrived from Siberia via the Bering land bridge, but new archeological discoveries indicate that the Clovis people may have been preceded by other humans thousands of years before.

No Clovis sites occur after the onset of the Younger Dryas cold climate period. The Clovis people disappeared at the same time that the megafauna became extinct. Humans had endured colder conditions than those of the Younger Dryas. Why did they disappear? Did a natural catastrophe eliminate the Clovis people along with the megafauna?

The Clovis people were replaced by a major wave of humans that populated America around 11,000 years ago, but their migration patterns are not well understood. The re-population of North America from humans already living in South America has been considered because sites occupied by humans 14,800 years ago have been found at Monte Verde in southern Chile. Tom Dillehay, an anthropologist from Vanderbilt University, and his colleagues have found additional material from fireplaces at this site with dates that cluster around 33,000 years ago, although these older dates have not been corroborated by other evidence.

The fact that humans had settled in South America before the corridor opened between the Cordilleran ice sheet and the Laurentian ice sheet means that the early settlers of the Americas may have arrived in boats from Asia following a route along the Pacific coast.

Dennis Stanford of the Smithsonian Institution has also suggested the Clovis people were not the first inhabitants of America, and that they were preceded by people of European origin who had come across the frozen Atlantic Ocean. Sites with charcoal from cooking fires, such as the Meadowcroft Rockshelter in Pennsylvania date from 13,200 to 19,200 years ago. The Cactus Hill site in Virginia has evidence of human habitation between 18,000 and 20,000 years ago. The bifacial stone tools with fluted bases found at the site have great similarity to those of the Solutrean flints from northern Spain and France. Stanford thinks that the Clovis culture emerged from Solutrean people who crossed the Atlantic during the ice age using boats similar to those made by the Inuits. During their trek along the edge of the ice on the Atlantic Ocean, these stone-age people fed on fish and sea mammals until they made their way to America.

DNA analysis of Native Americans has not produced convincing evidence of the Solutrean hypothesis. The genetic research indicates that the first colonizers of America came from Asia, although perhaps in several migrations.

It is hard to verify the settlement of America during the ice age. Many of the sites are now under water in the continental shelf because the level of the ocean has risen substantially since the human settlement. However, many Clovis-type sites have been found along the Mississippi River and its tributaries providing ample evidence that North America had a significant Clovis population before the passage from Beringia opened up.

Human societies tend to organize in similar ways. Hunter-gatherer groups had women, children and strong individuals who defended the camp from predators and enemies. Some of the less demanding

18

tasks like tending fires, cleaning skins to make clothing, flaking stone tools, gathering roots, bird eggs, nuts and fruits, and fishing could be assigned to the women, to the teenagers who still did not have the strength and stamina for hunting big game, and to the men who could not participate in the hunt because of illness, injuries or old age. The paleolithic people were modern humans who had not yet domesticated animals or cultivated crops. Their emotions and attitudes would have been similar to those of any modern culture. The survival of the group depended on being able to help each other.

Did the Clovis people have the capability of exterminating the megafauna? It is not likely. Hunting large animals was a dangerous activity that required highly organized groups with a lot of mobility and good hunting strategy. Mammoth hunters were probably nomadic because it would have been difficult or impossible to bring back a large kill back to a base camp to share with the women and children. When a mammoth was killed, it was probably better to relocate the camp temporarily to the kill site. This would have made it possible to defend the carcass from lions and other predators.

An excavation in Florida at the Page-Ladson site has found stone tools and mastodon bones that, based on radiocarbon dating, were deposited 14,550 years ago (Halligan, et al., 2016). The site is now under water, but at that time, it was next to a pond in a bedrock sinkhole within the Aucilla River. This is evidence that hunter-gatherers along the Gulf Coastal Plain coexisted and hunted the megafauna for 2000 years before the animals became extinct at approximately the onset of the Younger Dryas cooling event. The location of the find also provides some hints about the hunting methods using the river as a barrier. The hunters probably hid in vegetation along the banks of the river and attacked the mastodon knowing that the animal would try to escape by going back to dry land rather than risk going into deeper water. These hunters were predecessors of the Clovis people who used stone tools made by biface flaking, but without the symmetry and detail of the Clovis tools.

Many Clovis sites occur in a charred layer of soil. It is not known if the Clovis people used fire to clear the land or to divert animals for hunting. The use of intentional fires could explain the charred layers found at diverse Clovis sites. The fires would have reduced the habitat for the animals and forced them into smaller areas to facilitate hunting. Nomadic people would have been more likely to use fire as a hunting strategy because land without vegetation would not support herbivores and the hunters would have had to find new locations frequently. The carbonized layer has also been attributed to a high-temperature shockwave from an extraterrestrial airburst (Firestone, 2009).

The stone-age people would have known that fishing and gathering clams, oysters and snails was a less dangerous way to find food than trying to kill large animals. It would have been practical to establish camps close to the fishing areas and reinforce them against predators and the weather. Small game, rodents, rabbits, turtles, snakes and frogs could also have been hunted without straying too far from the camp and risking life and limb pursuing large animals and avoiding carnivorous cats.

Anthropologists Donald K. Grayson and David J. Meltzer reviewed material from 76 Clovis-age sites in 2002 and found that only 14 of those sites provide strong evidence for the hunting of mammoth and mastodon. These researchers concluded that there is no evidence to support the argument that humans played a significant role in causing the Pleistocene extinctions.

***

END OF THE ICE AGE

The last glacial period or ice age occurred during the last 100,000 years of the Pleistocene Epoch. The cover of ice over the land advanced and retreated several times. A prolonged warming period after the Last Glacial Maximum opened an ice-free corridor across Canada. The warming period, known as the Bolling-Allerod interstadial, started around 14,700 years ago and was interrupted abruptly approximately 12,900 years ago with the onset of the Younger Dryas cooling period.

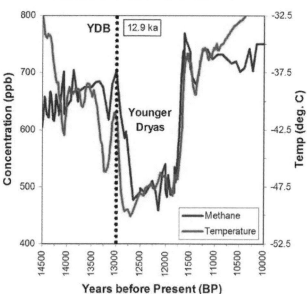

**Temperature drop during Younger Dryas
Modified from Firestone (2007)**

The Younger Dryas interval was first recognized in European pollen records that showed the reappearance of a cold-tolerant Arctic flowering plant (*Dryas octopetala*) from which the name of this period was derived. Temperatures in the Northern Hemisphere suddenly returned to near-glacial conditions, as confirmed from ice cores in Greenland and paleoclimate records from many parts of the world.

How do we know the past temperature of the Earth with such precision? The ice in the Greenland glaciers consists of layers that formed from each year's precipitation. Each layer of ice has chemical differences that provide a record of the Earth's temperature, and the records go back as far as 250,000 years ago. The ratio of oxygen-18 to the more common oxygen-16 in the molecules of water forming the ice varies with temperature in a predictable fashion. This makes it possible to determine the temperature of the Earth at a particular time by measuring the ratio of the oxygen isotopes in the corresponding layer of ice. Air bubbles trapped in the ice contain samples of the air at the time that the snow fell, so it is also possible to determine the past concentration of greenhouse gases like carbon dioxide or methane.

Scientists have wondered what caused the Earth's temperature to drop so suddenly and whether the sharp drop in temperature contributed to the megafauna extinctions. One popular theory about the cooling event proposes that the large amount of melt water from the glaciers altered the circulation pattern of the Atlantic Ocean and prevented the warm water of the tropics from traveling toward the American coast (Broecker, 2006).

Richard B. Firestone of the Lawrence Berkeley National Laboratory along with a long list of co-authors proposed in 2007 that one or more large, low-density extraterrestrial objects exploded over northern North America 12,900 years ago, partially destabilizing the Laurentide Ice Sheet and triggering the Younger Dryas cooling event. The energy of the impact also caused extensive fires and contributed to the megafaunal extinctions in North America.

As evidence of the impact, Firestone reported finding nanodiamonds that can only be produced at high pressures and small spherules that could have been produced from molten materials after an impact. Besides testing several Clovis sites, Firestone and his colleagues tested 15 Carolina Bays because previous researchers had suggested that the bays might be related to an extraterrestrial impact, although the mechanism by which the bays were formed was not known.

The extraterrestrial impact seemed to tie together the extinction of the megafauna and the Younger Dryas cooling event. However, the impact experts were not convinced. The "evidence" provided by Firestone and his colleagues was deemed to be insufficient. Firestone wrote another paper in 2009, this time by himself, providing additional details about the different types of spherules and other chemical indicators of an impact. He pointed out that, based on their orientations, the Carolina Bays appeared to radiate from the Great Lakes region or Hudson Bay thus supporting the impact theory, and he argued against the hypothesis that the Carolina Bays had been formed over thousands of years by strong winds, although he did not explain the mechanism by which the bays formed.

In 2011, Nicholas Pinter from Southern Illinois University along with a cadre of eminent academic scientists wrote a scathing "requiem" paper about the impact hypothesis proposed by Firestone. Pinter's paper refuted all of Firestone's evidence point-by-point and chastised him and his colleagues for not presenting "recognized and expected impact markers" and for proposing "impact processes that were novel, self-contradictory, rapidly changing, and sometimes defying the laws of physics."

The article by Pinter presented scientific arguments against the Younger Dryas Impact Hypothesis and the methodology used by its proponents. Pinter's paper listed overwhelming evidence that none of the impact markers presented by Firestone and his associates could be used reliably to prove that there had been an airburst or impact of an extraterrestrial object. The micrometeorite particles and extraterrestrial helium could not be substantiated. The report of increased iridium could not be reproduced. The Carolina Bays did not have meteoritic material and their dates were too diverse to have been caused by a single event. The nanodiamonds could have been deposited by various sedimentary processes and were not necessarily attributable to an extraterrestrial event. Finally, the microspherules could be fungal structures or termite dung. According to Pinter and his colleagues, the impact hypothesis was truly dead. The scientific community was warned to adhere to a rigorous approach using unambiguous criteria for extraterrestrial impacts and avoid being too eager to jump aboard the "impact bandwagon" when confronted with unusual geological evidence.

The requiem paper had a chilling effect on the scientific community. The Geological Society of America stopped accepting papers that tried to associate impacts to the Younger Dryas event or to the Carolina Bays. Research funding for these topics also vanished.

However, the impact hypothesis refused to die. In 2013, Michail I. Petaev and his colleagues from Harvard University published a paper announcing that they had found a platinum anomaly in the Greenland ice sheet that pointed to a cataclysm at the onset of Younger Dryas. The authors concluded that the combination of chemical compounds found in the Greenland glacier ice core hinted at an extraterrestrial source of platinum, possibly from an iron meteorite of low iridium content, and that such a meteorite was unlikely to result in an airburst or trigger wildfires over large areas of America. Such a meteorite would have passed through the atmosphere and impacted the ground. There seemed to be evidence of an extraterrestrial impact after all, although quite different from what Firestone had described.

The end of the ice age came shortly after the end of the Younger Dryas event and this marked the transition between the Pleistocene and Holocene epochs 11,400 years ago.

***

EXTRATERRESTRIAL IMPACTS

It was only in the middle of the 20th century that it became possible to distinguish unambiguously the characteristics of impact craters from volcanic craters. The study of Meteor Crater in Arizona, also known as Barringer Crater, played a key role in the development of the criteria; it was the first recognized impact crater on Earth. All extraterrestrial impacts are caused by meteorites because the term "meteorite" defines any celestial object, whether a comet or an asteroid, that strikes the ground after surviving its passage through the atmosphere.

Meteor Crater with a diameter of 1.2 km formed from the impact of an iron meteorite.

American geologist Eugene M. Shoemaker studied Meteor Crater extensively (Shoemaker, 1960). Cores taken from the bottom of the crater provided information about the layers underlying the cavity. The material surrounding the crater hinted at the explosive forces that had shaped the crater. The layers immediately around the rim of Meteor Crater are in the reverse order of the adjacent terrain. The inverted stratigraphy is a common characteristic of impacts. Microscopic examination of the minerals in the crater revealed shocked quartz (coesite), a form of quartz that has a unique crystal structure produced by intense pressure. The deformation features, visible under polarized light, consist of narrow planes arranged in different orientations than the grain's crystal structure. Shoemaker had encountered similar crystals at the Nevada test sites where atomic bombs had exploded.

Further study has shown that Meteor Crater was made approximately 50,000 years ago by the impact of an iron meteorite with a diameter of 40 meters, weighing about 300,000 metric tons and traveling at approximately 12 kilometers per second. The energy of the impact has been estimated as equivalent to 2.5 megatons of TNT.

Shoemaker established the field of planetary science through the meticulous observation of large impact structures and microscopic examination of the minerals that could be used as indicators of a crater's origin. The systematic analysis used by Shoemaker enabled scientists to distinguish volcanic craters from impact craters, even after significant structural alteration by wind or water erosion.

By the time that the Apollo astronauts had landed on the Moon, Shoemaker had established that, in addition to erosion and volcanism, impact cratering was one the most widespread geological processes occurring in the solar system. If every solid body had craters, it was impossible for the Earth to have escaped the impacts, although the craters might now be obliterated by erosion and volcanic processes.

In 1989, Prof. Jay Melosh published a book "Impact Cratering: A Geologic Process" that included everything that was known about extraterrestrial impacts and their mathematical analysis. A quarter century after its publication, the book is still widely referenced.

An extraterrestrial impact starts with a contact and compression stage where the projectile transfers its kinetic energy to the surface in an expanding hemispherical shock wave. The shock pressure exceeds the yield strength of the projectile and it will fragment and vaporize as soon as the pressure is released. In the excavation stage that follows, some target material is vaporized, melted or liquefied and displaced from the point of impact to form a bowl-shaped crater with a raised rim. The force of gravity takes over during the final modification stage and the ejected material falls in ballistic trajectories to the surface creating ejecta rays or inverted flaps. Molten minerals may collect at the bottom of the crater. The crater walls may become unstable and slide down, reducing the depth of the cavity.

The recognized and expected markers of an extraterrestrial impact consist of raised rims around the crater, meteorite fragments within or surrounding the crater, petrographic shock indicators such as crystals with planar deformation features (PDFs),

29

layers of fractured rocks (breccias) within the crater, and shatter cones in the surrounding rocks which are aggregates of mineral crystals fused together in conical shapes by the passage of an intense shock wave. When meteorite fragments cannot be found, the proposed impact site should at least contain traces of elements that are more abundant in space rocks than in the Earth, i.e., siderophile elements, such as iridium.

From this perspective, we can see why the community of experts dismissed Firestone's impact extinction hypothesis. Even if Firestone was right about the extraterrestrial event, his evidence appeared to be circumstantial and not based on the precise science that had developed over the previous 60 years. Firestone had not shown the location of an impact site or crater, and his proposal of an explosion of a low-density extraterrestrial object over the Laurentide Ice Sheet seemed purely hypothetical without any physical validation. In addition, the nanodiamonds and microspherules that he presented as evidence had not been established as infallible proof of an impact under the strict standards of hypervelocity impact science.

The possibility that an impact event could be proven was renewed with the discovery of an elevated amount of platinum by Petaev in the Greenland ice corresponding to the onset of the Younger Dryas event and by the confirmation by James H. Wittke from Northern Arizona University and 27 other scientists that "impact spherules" had been found across four continents in soil layers corresponding to the Younger Dryas event.

Perhaps the impact hypothesis needed to be approached from a new point of view. The Carolina Bays seemed to radiate from the Great Lakes region where the impact was supposed to have happened, but the mechanism for the formation of the bays was still not known. Could the bays be used to prove that there had been an extraterrestrial impact and shed some light on the mystery of the megafauna extinction?

***

THE CAROLINA BAYS

The Carolina Bays are shallow marshy depressions with elevated sandy rims that are filled to a significant extent by sand and silt. Since early colonial days, farmers have drained some of the bays and planted crops on them to take advantage of their particularly fertile soil.

The Carolina Bays are so large that it is impossible to determine their shape from ground level. It was only in the 1930's with the advent of aerial photography that the bays attracted attention because of their regular elliptical shape and common alignment. Bays have been found all along the Atlantic coast from southern New Jersey to northeastern Florida.

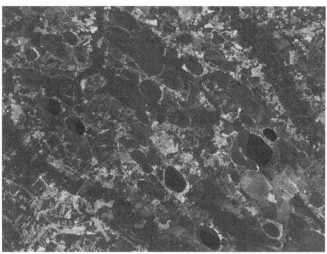

**Aerial view of Carolina Bays
near Elizabethtown, NC**

The Carolina Bays have very specific morphological characteristics. The bays in the East Coast have a basic elliptical shape with northwest-southeast alignment and a thickened southeast rim. The major axes of the ellipses vary from 60 meters to 11 kilometers (about 200 feet to 6.8 miles). Deviations from the northwest-southeast orientation appear to be systematic by latitude. The northernmost bays are oriented toward the northwest, whereas those further south have their major axis toward the north. The bays are shallow depressions with a maximum depth of about 15 meters (49 feet). Large bays tend to be deeper than small bays, but the deepest portion of any bay is offset to the southeast from the bay center. The elevated bay rims may be as high as 7 meters.

Carolina Bays frequently overlap other bays without destroying the morphology of either depression. One or more small bays can be completely contained in a larger bay, and the ground layers beneath the bays do not seem to be distorted.

Bays occur only in unconsolidated sediments and in loose sandy soil away from modern river flood plains and beaches, but the bays may occur at different topographic levels. Bays are either filled or partly filled with silt of organic and inorganic origin. Many Carolina Bays have been destroyed partially or totally by farming or urbanization.

Melton and Schriever of the University of Oklahoma proposed in 1933 that the Carolina Bays could be the scars of impacts from a meteorite shower or colliding comet that must have come from the northwest to produce the observed alignment of the bays. These researchers observed that the bays have a narrow range of width-to-length ratios, which they defined in terms of ellipticity (length minus width, divided by the length).

Since the bays are very large, it was expected that the meteorites that created them would also be large, but no trace of meteorites or extraterrestrial material could be found in or around the bays. Eventually, all the theories that the bays were created by asteroids or fragments of exploding comets were discarded because the bays do not have any of the characteristics of hyperspeed impacts.

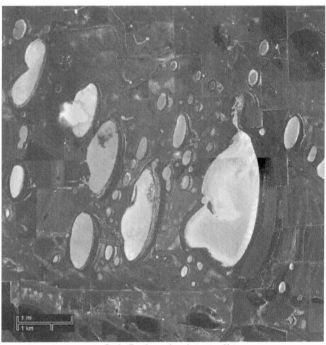

Salt Lakes in Australia

Geologists started proposing ideas of how wind and water processes could have made the Carolina Bays as an alternative to impacts. Models of the oval salt lakes in Australia which form when a marshy area starts to dry up have been considered, but the process does not create raised rims. The comparisons of the Carolina Bays to the lakes that form in Alaska, when permafrost thaws, also fail because the lakes are not all perfect ellipses, they have no overlaps, and they do not have raised rims or preferential rim thickening. These thermokarst lakes are generally aligned in the direction in which water flows to lower levels.

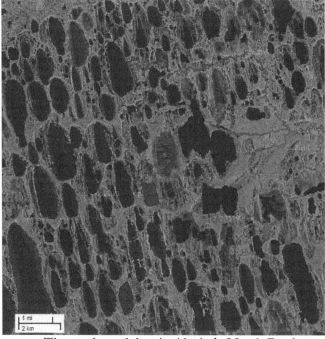

Thermokarst lakes in Alaska's North Bank

Prof. D. Johnson in 1942 proposed that the bays had originated as shallow basins created by upwelling springs, and that wave action and wind had created the sandy ridges. Many subsequent hypotheses by other researchers tried to describe how terrestrial processes could have created the bays, but they never explained satisfactorily the elliptical shapes and the regular orientation.

In 1954, C.W. Cooke of the Geological Survey, proposed that the forces that shape marine eddies depend on the latitude and that this could account for the elliptical shape and axial orientation of the Carolina Bays. The author had a disclaimer that his conclusions were purely deductive without experimental or mathematical verification, so his suggestion did not garner much support.

Another hypothesis of bay formation stated that marine waves and currents could produce depressions that are later modified by wind or ice-push processes to create the raised rims (May and Warne, 1999). This hypothesis proved inadequate because many bays are on ground that is too high above sea level to have been affected by marine currents or too far south to have had the cold temperatures needed for ice-push processes.

Analysis of the Carolina Bays has produced a wide range of dates using Optically Stimulated Luminescence (OSL). This has caused scientists to reject the idea that all the bays were created simultaneously by a single impact event. Mark J. Brooks from the University of South Carolina and his colleagues have proposed that there were multiple periods, separated by thousands of years, during which the Carolina Bays formed by various wind and water processes (Brooks, 2010).

The main deficiency of the theories of bay formation by terrestrial processes is that the mechanism by which wind and water currents could carve perfect ellipses with raised rims and a common alignment has never been observed or demonstrated. If we add the requirement that the bay formation process should explain how to create overlapping bays and the specific width-to-length ratios, it becomes evident that the existing theories are inadequate and that they only survive for lack of better explanations.

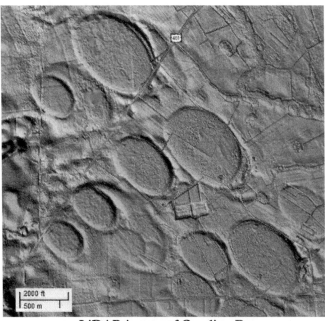

LiDAR image of Carolina Bays
near Bowmore, North Carolina

In the early 1960's LiDAR was invented. It combined laser imaging with radar's ability to calculate distances. When LiDAR was used on the East Coast of the United States, thousands upon thousands of Carolina Bays were discovered that were not visible in ordinary aerial surveys. In some regions, it was impossible to find a piece of land that was not covered by bays. Prouty (1952), using images obtained by aerial photography, had estimated that there were about half a million Carolina Bays, but the LiDAR images indicated a greater number.

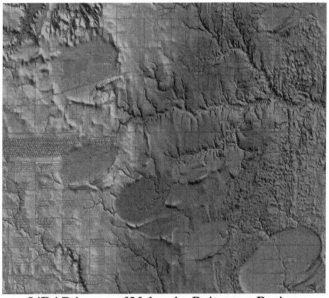

LiDAR image of Nebraska Rainwater Basins

LiDAR and similar technologies that can measure small differences in the elevation of surface features are very useful in civil engineering and in geology.

Zanner and Kuzila reported in 2001 that Nebraska had elliptical depressions with the same characteristics as the Carolina Bays. These elliptical features, which also had width-to-length ratios equal to the Carolina Bays, were called "Nebraska rainwater basins." However, in Nebraska the ellipses were oriented from the northeast to the southwest, almost perpendicular to the alignment of the Carolina Bays. The Nebraska bays were highly eroded and covered with loess (windblown dust). Unlike the Carolina Bays, the elliptical features could not be detected visually from airplanes or satellite images because they were well camouflaged by cornfields and farms. Only technology like LiDAR could see them.

The alignment of the rainwater basins made it evident that the elliptical bays in the East Coast and in Nebraska seemed to radiate from a point in the Great Lakes region. The changes in the orientation by latitude of the Carolina Bays now made sense. Only a common mechanism could have formed ellipses with the same width-to-length ratios in the East Coast and in Nebraska radiating from a common point by the Great Lakes. The proposals that water mechanisms could have created the Carolina Bays over the last 100,000 years were now invalid because similar processes could not have created the Nebraska rainwater basins. Part of Nebraska had been under the Western Interior Seaway during the Paleocene Epoch, but that inland sea drained away 60 million years ago when the Laramide orogeny elevated the land and started forming the Rocky Mountains. The Nebraska rainwater basins formed at elevations of 600 to 700 meters above sea level and far from any oceans.

When we compare the elliptical shapes of the Carolina Bays with the irregular shapes of the thermokarst lakes in Alaska or the salt lakes in Australia, we must conclude that they were created by different processes. Even if some of the Alaskan or Australian lakes are elliptical, it can be shown statistically that they are very different from Carolina Bays. Examine the image of the thermokarst lakes, for example, and count the proportion of elliptical lakes. Let us say that one out of thirty is close to elliptical. The probability of finding ten adjacent elliptical lakes would be $(1/30)^{10}$ or 1.7×10^{-15}, which is highly unlikely. It would be as unusual as flipping a coin and getting 47 heads in a row. However, it is very common to find ten adjacent elliptical Carolina Bays, so they must have been created by a different mechanism than the lakes. The mechanism that produced the Carolina Bays had to have a high probability of creating elliptical conic sections.

In the middle of the twentieth century, doing research on the Carolina Bays had been a laborious process that required getting aerial photographs and geological surveys from specialized sources. Google Earth changed all that. Now, anyone with a computer and access to the Internet could examine detailed maps of almost any place on the Earth, and the views could be overlaid with LiDAR images. Using this enhanced technology, Davias and Gilbride used great circle trajectories to extend the axial orientations of the ellipses and made adjustments for the rotation of the Earth to compensate for the Coriolis Effect. In 2010, they found that Saginaw Bay in Michigan was the focal point from which the Carolina Bays and the Nebraska rainwater basins radiated. Previous researchers had just used straight lines on flat maps, and they had not taken into consideration the time of flight of projectiles hitting a rotating sphere like the Earth.

The early proponents of impact hypotheses, such as Melton and Shriever or Prouty, did not know about the Nebraska rainwater basins, so they had assumed that the projectiles that made the Carolina Bays had come from outer space. Knowing that the projectiles originated from Saginaw Bay in Michigan made it possible to consider that the projectiles consisted of ice ejected from the Laurentide ice sheet that covered Michigan during the late Pleistocene Epoch. The mechanism of how this could have happened is explained in the following chapter.

***

THE GLACIER ICE IMPACT HYPOTHESIS

The Carolina Bays present an enigma. Their alignment toward the Great Lakes region is what would be expected of an impact event. However, the bays do not have any of the features accepted as evidence for extraterrestrial impacts like meteorite fragments or mineral crystals deformed by the extreme pressure of a cosmic impact.

The Glacier Ice Impact Hypothesis (Zamora, 2013) reconciles this apparent incompatibility by proposing that the Carolina Bays were made by impacts of terrestrial ice at ballistic speeds slower than those of asteroids or comets. An extraterrestrial impact on the Laurentide ice sheet at the intersection point in Michigan determined by Davias would have ejected ice boulders, and the secondary impacts by the ejected ice could have created the Carolina Bays. In his 2009 paper, Firestone justified the absence of a crater for the extraterrestrial impact by explaining that an impact on an ice sheet would have absorbed much of the energy and that most of the ejecta would have been ice. Unfortunately, he did not elaborate about the mechanics of the ejecta.

The Glacier Ice Impact Hypothesis uses the constraints imposed by the laws of physics to determine the physical and dynamic attributes of the ejected ice boulders that are necessary to create the Carolina Bays. These are the four main premises of the hypothesis that will be examined in greater detail below:

1. An asteroid or comet impact on the Laurentide ice sheet ejected glacier ice boulders of various sizes.
2. Seismic shock waves from the impacts liquefied unconsolidated ground close to the water table.
3. Oblique impacts of glacier ice boulders on liquefied viscous ground created slanted conical craters.
4. Viscous relaxation reduced the depth of the conical cavities and produced shallow elliptical depressions.

The reason why the Carolina Bays have not been recognized as impacts before is because the books about impacts, such as those by Melosh (1989) or French (1998) focus on extraterrestrial high-speed impacts and do not cover the ordinary terrestrial impacts that a child throwing stones at a mud puddle might encounter. These terrestrial impacts on viscous surfaces are not novel. They just have not been interesting enough to put in reference books. More attention has been paid to the formation of rampart craters on Mars (e.g., Greeley, 1980) than to the mechanics required to produce elliptical craters on Earth. To understand the formation of the Carolina Bays, it is necessary to examine the behavior of impacts on viscous and plastic surfaces.

1. Impact on the Laurentide ice sheet

The first premise of the Glacier Ice Impact Hypothesis is that an asteroid or comet impact on the Laurentide ice sheet ejected glacier ice boulders.

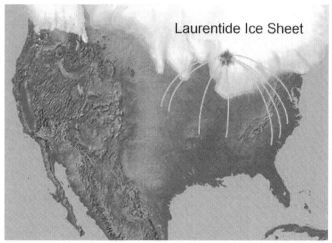

Laurentide Ice Sheet

**Extraterrestrial impact
on the Laurentide Ice Sheet**

Although no impact site has been found, the trace of platinum that Petaev found at the Younger Dryas boundary in the Greenland ice supports the idea that an extraterrestrial object could have had the chemical composition to survive its passage through the atmosphere and actually hit the ice sheet instead of just disintegrating in an airburst.

What happens when an extraterrestrial object hits a glacier? We know that an impact on a rocky surface generates a lot of heat, and melts and ejects pieces of rock. Something similar happens with ice.

Ice is brittle and a bad conductor of heat. Peter H. Schultz from Brown University conducted experiments with NASA's Ames Vertical Gun showing that ice shatters upon impact. Pieces of ice are ejected at high velocity in ballistic trajectories radiating from the impact site. This radial pattern is characteristic of impacts and could be responsible for the alignment pattern of the Carolina Bays and Nebraska Rainwater Basins toward the Great Lakes region. Stickle and Schultz (2012) have also shown that a layer of ice shields the underlying terrain from the effect of a hyperspeed impact and reduces subsurface damage.

Impact on an ice sheet ejects pieces of ice (Schultz 2009)

A layer of ice two kilometers thick would not offer as much resistance as rock to a meteorite during the contact and compression stage, but the shock wave would cause the ice to break up. The heat from the impact would produce great quantities of water and steam at great pressure, which would help to propel the ice pieces during the excavation stage. The ejecta produced by the impact would be mainly ice, water and steam. During the modification stage, when the energy of the impact has been dissipated and gravity

45

becomes the dominant force, a large quantity of water would flood the impact site, perhaps even washing away evidence of the impact. Over time, the water would drain under the remaining glaciers. The ice boulders ejected during the impact would travel above the atmosphere, as will be shown later, and blanket an area with a radius of about 1500 kilometers. Any water ejected above the atmosphere would turn into ice crystals in the vacuum of space and would block the light of the sun.

2. Soil liquefaction by seismic shock waves

The second premise of the Glacier Ice Impact Hypothesis is that seismic shock waves from the extraterrestrial impact and the ejected ice boulders liquefied the sandy soil along the east coast of the United States and in Nebraska along what once were the silt banks of the South Platte River. All Carolina Bays along the Eastern Seaboard and in Nebraska occur only on unconsolidated soil close to the water table. There are no bays on hard ground.

Sandy soil can be liquefied by vibrations when there is water near the surface. The vibrations suspend the sand grains in the water and reduce friction causing the soil to flow like a liquid. Soil liquefaction has been observed in earthquake zones like Japan and New Zealand. Cars and buildings may be submerged in the quicksand that forms when the ground vibrates. Earthquakes of magnitude 6 or higher are strong enough to liquefy saturated soil. The earthquake in Niigata, Japan that toppled buildings had a magnitude of approximately 7.5. A report from the U.S. Geological Survey by Eimers

(2001) shows that the areas of North Carolina that contain Carolina Bays have the water table within 1.5 meters (5 feet) from the surface. These areas would be susceptible to liquefaction today and probably also at the time that the bays formed.

It is important to establish that the ground where the Carolina Bays formed could have been liquefied because this made it possible for the ejected ice boulders to penetrate the surface. Ice boulders that hit rocky or hard terrain would have shattered without leaving any lasting marks on the landscape.

Buildings toppled when the soil liquefied during the 1964 earthquake in Niigata, Japan

A ground impact by a comet or asteroid in the Great Lakes Region would have generated seismic shock waves traveling at 5 km/sec that would have reached the Eastern seaboard approximately 3.5 to 4.5 minutes after the extraterrestrial impact. The shock waves from the primary impact might have had enough energy to liquefy saturated soil, but the high-energy impacts of the ejected ice boulders arriving 3 to 6 minutes later would have definitely liquefied the surface. The multitude of impacts of huge ice boulders would have caused seismic vibrations to maintain the surface liquefied for the duration of the bombardment by the glacier ice chunks.

3. Oblique impacts create slanted conical cavities

Craters made by extraterrestrial impacts are generally bowl-shaped. This happens because the projectile is destroyed upon contact and most of its kinetic energy is transferred to the point of impact. The resulting hemispherical shock wave creates a bowl-shaped crater. Even moderately oblique impacts at high speeds create hemispherical shock waves and bowl-shaped craters because the transfer of kinetic energy is virtually instantaneous. Scientists have considered that impacts create conical craters and that the modification by melting and collapse of the crater walls is what creates the bowl shape. The Linne crater on the Moon has been described as being conical and thus a suitable archetype of a pristine crater that has not been degraded by in-filling (Garvin, 2011).

Conical craters are also made on viscous targets when the projectile is not destroyed by the impact. The excavation phase starts as the projectile continues to travel through the viscous medium creating a conical shock wave until it is stopped by friction. A viscous surface with low elasticity will retain the conical shape, and the cavity will gradually be modified by gravity through viscous relaxation.

The third premise of the Glacier Ice Impact Hypothesis is that oblique impacts of glacier ice boulders created slanted conical cavities that were later modified by gravity to form elliptical bays. Experiments demonstrate that conical cavities are created by impacts of ice spheres on a mixture of clay and sand mixed with enough water to have the consistency of mortar. Oblique impacts produce slanted conical cavities that are analogous to geometrical conic sections.

Conical cavity created by oblique impact

The experimental oblique impacts demonstrate that an overturned flap is formed as the projectile penetrates the viscous surface. The overturned flap becomes a raised rim surrounding the elliptical cavity. Oblique impacts tend to push surface material in the direction of the impact thereby creating a thickened rim at the terminal end of the cavity.

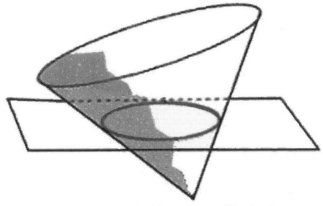

A tilted conical cavity forms an elliptical crater

Oblique impacts on viscous surfaces can explain three of the characteristics of the Carolina Bays: the elliptical shape, the raised rims surrounding the cavity, and the thickened rim at one end. In addition, if the projectiles originate from a common point, the alignment of the bays can also be explained.

Overturned flaps become raised rims

Wind or water processes cannot guarantee the

formation of perfect ellipses with raised rims, whereas these structures result naturally from oblique impacts on viscous surfaces. The fact that the elliptical shape of the bays corresponds to geometrical conic sections makes it possible to use mathematics to calculate the original depth of the bays and the location where the ice boulders might have ended up. Future exploration of the Carolina Bays could use this information to find stones that might have been carried within the glacier chunks.

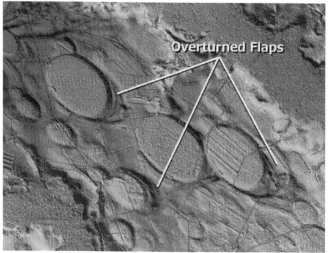

Examples of overturned flaps in Carolina Bays

The impact angle θ may be estimated from the ratio of the minor axis (**W**) to the major axis (**L**) using the relation **sin(θ) = W/L**. The ellipticity of an ellipse is defined as the length minus the width, divided by the length. Thus, **ellipticity = 1 − W/L**. In their 1933 paper, Melton and Schriever measured the ellipticities of dozens of Carolina Bays and calculated that they could have been made by impact angles varying from 35 to 55 degrees.

Bay shape corresponds to angle of impact

Melton and Schriever also plotted the ellipticity of the bays against their length and noticed that the bigger bays have greater ellipticity. This was confirmed by Prouty in 1952. In general, small bays are more circular than the larger bays, which have a more elongated shape. This observation can be explained by the atmospheric drag on the ice boulders during re-entry from their suborbital space flights. Melosh (1989) has equations showing that the decrease in velocity due to atmospheric drag is inversely proportional to the mass of the projectile. This means that smaller projectiles would be slowed more by the atmosphere and, therefore, hit the ground at a more vertical angle than the larger projectiles.

4. Viscous relaxation reduces cavity depth

Two arguments that have been used against the impact origin of the Carolina Bays are: 1) impacts would have disturbed the soil under the bays but the ground under the bays appears undisturbed and 2) the dates of the bays vary widely indicating that they must have originated at different times.

The fourth premise of the Glacier Ice Impact Hypothesis is that viscous relaxation reduced the depth of the cavities and produced shallow elliptical depressions that also restore the stratigraphy of the underlying terrain. This mechanism provides an explanation to counter both of the above arguments.

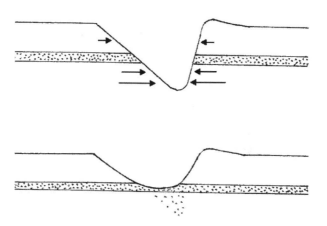

Viscous relaxation reverses the sequence in which the cavity was made.

Projectiles at ballistic speeds penetrate viscous surfaces by parting the existing layers of the target during the excavation phase of the impact. All the energy of the projectile is spent moving the material along its path. This process of plastic deformation is different from the impacts on hard surfaces in which the excavation phase actually disrupts the underlying layers.

The modification phase is driven by gravity. A cavity in a viscous surface is filled by flow of the material surrounding the deepest part of the cavity. This happens because pressure increases with depth forcing the material at the bottom to flow faster. The centripetal lateral flow of material starting from the bottom reverses the sequence in which the cavity was made and restores the stratigraphy. The cavity stops being filled when the pressure driving the flow cannot overcome the friction of the medium, so the final configuration of the cavity is a shallow depression analogous to the Carolina Bays.

The reconstitution of soil layers by viscous relaxation is an unfamiliar concept because it does not occur in cavities made by impacts on hard surfaces due to the destructive nature of the impact. By contrast, an impact on a viscous surface is just a plastic deformation that parts the material and does not change the chemical or physical composition of the medium. The following images show a cavity made by the oblique impact of an ice ball on a sand-clay target with an underlying red layer approximately two centimeters below the surface. The penetration of an ice projectile through the red layer drags along some of the red material. The depth of the cavity is gradually reduced by viscous relaxation, but the red layer remains at the same level.

Oblique impact in surface with red layer

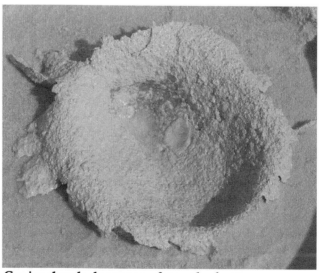

Cavity depth decreases from the bottom up

The red layer remains at the same level

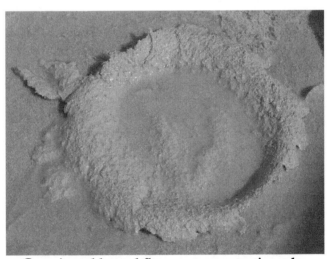

Centripetal lateral flow restores stratigraphy

The wide range of ages obtained for the Carolina Bays has been used to suggest that the bays formed in multiple episodes over a long period of time. In 2010, Mark J. Brooks and his colleagues from the University of South Carolina reported that, based on 45 dates obtained by Optically Stimulated Luminescence (OSL), the wind processes that modified the shorelines of the bays occurred in five stages dating from 12,000 to 140,000 years ago. This presumably confirms that the bays are not features that were created by a single impact event.

How does OSL dating work? Cosmic rays and ionizing radiation from naturally occurring radioactive elements in the earth can cause electrons to become trapped in the crystal structures of buried quartz and other minerals. OSL is able to free the trapped electrons and produce luminescence in proportion to how long the quartz has been buried. Exposure of quartz to sunlight eliminates the trapped electrons and resets the clock of the luminescence signal. In essence, OSL dating estimates the time since last exposure to sunlight for quartz sand and similar materials. Samples for OSL dating have to be taken in soil not exposed to light and they are processed in rooms with low-intensity sodium vapor lights that emit a single wavelength (589 nm) and are adjusted to provide enough sensitivity for the human eye but not bleach the samples.

The only way to reconcile the great span of dates of the Carolina Bays is to note that an ice projectile parts the target material without mixing it, and only the new surface area of the conical cavity is exposed to light. When the cavity depth is decreasing by viscous relaxation, no additional material is exposed to light. After the bay reaches its final configuration, the bay has the same underlying physical characteristics as before the impact. The material immediately below the bay surface has not been exposed to light throughout the bay formation process and its date will be the date of the original target terrain and not the date when the bay was formed.

Some adjacent bays separated by less than one kilometer have been reported to differ in OSL age by about 10,000 years (Brooks 2010, Fig. 4). However, if OSL can only determine the age of the target surface, then it is not a valid method for establishing the time when the bays were formed, and it becomes possible to postulate that all the bays could have been created contemporaneously. Understanding the mechanism by which the Carolina Bays were formed enables us to exclude OSL as a valid method for dating the creation of the bays.

Another consideration regarding the use of OSL is that if the impacts that made the bays occurred at night or under a dark cloud cover produced by the extraterrestrial impact, even the conical surfaces of the impacts would not have been exposed to sunlight at the time of their formation.

Overlapping Bays

A curious characteristic of the Carolina Bays is that many of them overlap while maintaining their regular elliptical shape. Typically, large bays may have smaller bays within them or they may intersect portions of adjacent bays.

Wind and water processes cannot explain the

mechanism by which perfectly elliptical overlapping bays are created, but adjacent oblique conical impacts transform into overlapping bays after undergoing viscous relaxation, as illustrated by the following experimental impacts. A bay that covers another one is created later in time, thus providing a chronology of stratigraphic formation.

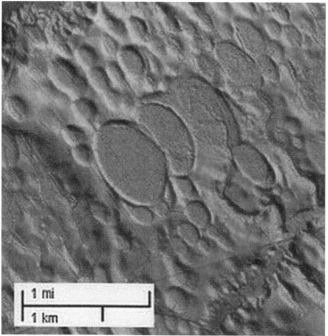

Overlapping Bays

Bays that overlap may produce different patterns, including side-by-side, one bay within another one, or heart-shaped bays formed by two projectiles coming from different directions due to collisions within the ejecta curtain.

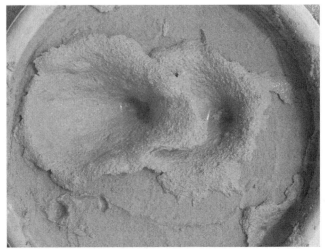

Adjacent conical craters

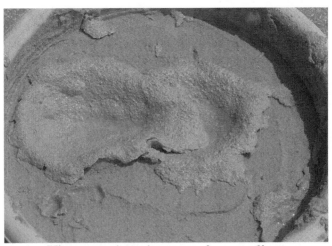

**Viscous relaxation transforms adjacent
conical impacts into overlapping bays**

**The overlaps indicate
the sequence of the impacts**

Impact mechanisms explain the elliptical shape of the Carolina Bays better than wind and water processes. The eolian and lacustrine mechanisms cannot explain the radial orientation toward the Great Lakes of the bays throughout the East Coast and Nebraska at different elevations in different geologic settings, and there is no mathematical or experimental model that shows how wind and water processes can create elliptical bays with specific width-to-length aspect ratios. By contrast, the interpretation of the Carolina Bays and Nebraska rainwater basins as conic sections whose width-to-length ratio corresponds to the angle of impact is consistent with their geomorphology and provides a mathematical foundation for their analysis.

***

THE COMET IMPACT

Oblique impacts on viscous surfaces provide a model that explains many of the characteristics of the Carolina Bays. Therefore, it is reasonable to propose that the bays were formed by secondary impacts of ice boulders ejected by an extraterrestrial impact on the Laurentide ice sheet that covered North America. After the ice boulders melted, no trace was left of the projectiles, except, perhaps, for some glacier rocks that might have been imbedded within them.

Up to now, the site of the extraterrestrial impact has not been found and it is not known whether a comet or an asteroid hit the Laurentide ice sheet. A comet is like a dirty snowball containing frozen gases and rocks that originate beyond the orbit of Jupiter. Comets travel at speeds of 45 to 50 km/sec. Asteroids come from the region between Jupiter and Mars. They are rocks or chunks of metal, usually iron-nickel alloys that travel at around 15 km/sec. An impact by a comet is much more destructive than an impact by an asteroid of similar size because of its greater kinetic energy.

The evidence for an impact has been the platinum anomaly found by Petaev and the various types of spherules reported by Firestone and Wittke. The chemical composition of the extraterrestrial projectile is not known. The analysis by Petaev and his colleagues suggest an iron meteorite with high platinum content, but the spherules that have been found at widely distributed places have compositions similar terrestrial rocks. Metallic elements that may be of extraterrestrial origin, like iridium, at the Younger

Dryas boundary have been harder to confirm. If the projectile had been a large metallic asteroid, it would have left a substantial residue of extraterrestrial elements instead of just trace amounts. It is more likely that a comet like the ones that filled the Earth's oceans caused the impact. A comet consisting mostly of water ice and rocky material could have crashed into the glacier leaving few traces that could be recognized using the established criteria for extraterrestrial impacts.

Impact and ejecta

Most of the Clovis artifacts and megafauna bones are found below a brown layer of dirt containing carbonized plant remains called the Bolling-Allerod paleosol. The carbonized layer occurs in many sites at the Younger Dryas level throughout the United States. This is the reason why it is thought that the flash of the comet impact must have started wildfires in a large area of the United States. A comet impact traversing the atmosphere at 45 km/sec would have crossed the atmospheric layer in a couple of seconds and produced an enormous fireball. The spike of thermal radiation could have ignited fires far away from the impact zone, just like it happens in nuclear explosions, although, as mentioned before, it is possible that the carbonized remains are from fires that the Clovis people used as a hunting strategy. An asteroid traveling at 15 km/sec would have needed about 7 seconds to penetrate the atmosphere, but it also would have radiated enough energy to ignite fires over a wide radius.

The projections of the major axes of the Carolina

Bays intersect by the Great Lakes, around Michigan. This would be the point of the extraterrestrial impact. When the comet hit the Laurentide ice sheet, the shock wave from the contact fractured the ice for many kilometers around the point of impact because ice is brittle and undergoes sudden material collapse at high strain rates.

If the comet nucleus penetrated to the bottom of the thick ice sheet during the contact and compression stage, the heat generated by the enormous transfer of energy would have melted large quantities of glacier ice and converted it to superheated steam. During the excavation stage, which lasted for about 30 seconds, the impact would have launched glacier ice fragments powered by the explosive power of the compressed steam. The ejecta would have also included water, mud and rocks from the saturated landscape under the ice sheet. The spherules that have been found throughout several continents were probably formed during this stage.

As discussed before, the impact angle at which the Carolina Bays were formed can be estimated from their width-to-length ratio. Many of the Carolina Bays were formed from impacts at an inclination of approximately 35 degrees from the horizontal. This means that the ejecta curtain from the extraterrestrial impact had a broad conical shape that expanded with time as the ice boulders followed ballistic trajectories from their launch position. The relatively low angle of the trajectories may be due to the rapid vapor plume expansion when ice was converted to steam at the point of impact.

We can calculate that an ice boulder ejected at an

angle θ of 35 degrees from Michigan to the South Carolina seashore would require a launch speed **v** of approximately 3.6 km/sec to cover the distance **D** of 1,220 kilometers using the following ballistic equation, where **g** is the acceleration of gravity.

$$D = (v^2/g)\sin(2\theta)$$

At this speed, which is almost eleven times the speed of sound, the ice boulder would reach its target about 7 minutes after launch. The trajectory would take the boulder 213 kilometers above the surface of the Earth, which is more than a hundred kilometers above the atmosphere. The time of flight **T** and the maximum height **H** of the trajectories are given by the equations:

$$T = (2v/g)\sin(\theta)$$
$$H = v^2\sin^2(\theta)/2g$$

Depending on the distances and launch angles, some ice chunks would have traveled at 3 to 4 km/sec and reached heights of 150 to 390 kilometers above the surface of the Earth.

In the vacuum of space, water cannot exist in liquid form. Any liquid water or slurries ejected above the atmosphere would have immediately created clouds of ice crystals. The water crystallization phenomenon is well known to astronauts when they vent their urine into space. The following is part of a 1976 transcript of astronaut Russel Schweickart talking to Peter Warshall about waste disposal in space:

Schweickart: Well, actually, in Skylab we did something similar to that. But on Apollo the urine then would go outside, and you'd have to heat the nozzle because, of course, it instantly flashes into ice crystals. And, in fact, I told Stewart this, the most beautiful sight in orbit, or one of the most beautiful sights, is a urine dump at sunset, because as the stuff comes out and as it hits the exit nozzle it instantly flashes into ten million little ice crystals which go out almost in a hemisphere, because, you know, you're exiting into essentially a perfect vacuum, and so the stuff goes in every direction, and all radially out from the spacecraft at relatively high velocity. It's surprising, and it's an incredible stream of ... just a spray of sparklers almost. It's really a spectacular sight. At any rate that's the urine system on Apollo.

The transformation of water into ice crystals may help explain what caused the onset of the Younger Dryas event. If the comet impact ejected large amounts of liquid water into space, the resulting ice crystals in low Earth orbit could have blocked the light of the sun for many years thus leading to a period of abnormal cold and darkness.

Ice chunk sizes

Knowing the speed at which the glacier ice was launched, we can estimate the size of the chunks of ice that made the Carolina Bays. The University of Arizona calculator for computing projectile size from crater diameter (Melosh 1999) allows the input of the crater diameter and various initial conditions to determine the diameter of the impacting object.

According to the Melosh equations, a spherical ice boulder with a diameter of 180 meters traveling at a speed of 3 km/sec would create a crater of 1-kilometer diameter when impacting at an angle of 45 degrees on a sandy target. This size is very common for Carolina Bays.

Using the formula for the volume of a sphere, the ice boulder would have a volume of 3 million cubic meters and would weigh 2.8 million metric tons. The ice boulder would be about the size of Yankee Stadium in New York City. The impact of such a huge boulder would have the energy of 1.27×10^{16} joules or 3.03 megatons of TNT. This is about 200 times more powerful than the bomb dropped on Hiroshima, which had a yield of 15 kilotons. This energy is comparable to an earthquake of magnitude 7.54, which can easily liquefying the soil.

If we consider that there are approximately 500,000 Carolina Bays, then the total energy of the impacts was 6.35×10^{21} joules or 1.5 teratons of TNT, which is 100 million times greater than the Hiroshima bomb. The energy of the comet impact must have been enormous because additional energy would have been converted to heat and seismic shocks.

The amount of ice ejected from the Laurentide ice sheet to form all the bays would have been approximately 1.5×10^{12} cubic meters of ice. This would correspond to an ice sheet with a thickness of 1 kilometer and a circular area with a diameter of 44 kilometers.

The comet impact probably ejected the glacier ice

in a circular pattern, as is the case for most extraterrestrial impacts. We only have evidence of the impacts that formed the Carolina Bays and the Nebraska rainwater basins, but many ice boulders must have fallen on ground that was too hard to form bays. This means that the amount of ice calculated based on the number of bays may substantially underestimate the actual amount of ice ejected.

The brittle ice boulders that fell on hard ground would have shattered on impact. The ice shards from the fragmenting boulders would have dissipated their energy by knocking anything in their path. An ejecta blanket containing 1.5×10^{12} cubic meters of ice distributed in a semicircle with a radius of 1500 km would have covered approximately half of the contiguous United States, including the Midwest and the Atlantic coast, with pieces of ice to a depth of about half a meter. However, since the estimates are based on the number of Carolina Bays, but they occur only on 16 to 20 percent of that semicircle, the ice ejected may have been at least 5 times greater, which means that the cover of crushed ice could have had a depth of 2.5 meters. Normally, the layer of ejecta is thicker near the impact point than further away.

It is possible that the onset of the Younger Dryas event was also influenced by the destruction of the forests and the new cover of crushed ice. The increased ice cover would have reflected the light of the sun back into space. The combination of greater flow of melt water from the impact point, orbiting ice crystals, and increased albedo from the ice cover could have certainly played a role in starting a period of cold climatic conditions.

Some day the exact location of the comet impact may be located by projecting the axes of the larger bays to the point of origin. The trajectories of the larger ice chunks would not have been easily altered during mid-air collisions within the ejecta curtain.

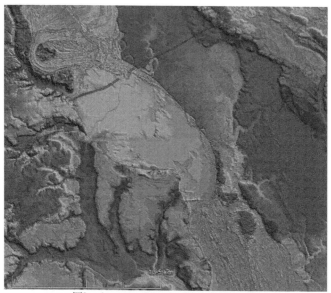

The largest bay in South Carolina
(Lat. 34.2542, Lon. -79.6126)

The largest bay found so far is situated near Florence, South Carolina. It has a major axis of 16.2 km and a minor axis of 7.7 km. The ice chunk that made this bay would have measured approximately 4 kilometers in diameter. Erosion and terrain movement have degraded the elliptical structure substantially, but it is recognized as a Carolina Bay because the orientation of the major axis and the ratio of the axes are consistent with other nearby bays.

The Carolina Bays are rapidly disappearing through erosion and urbanization. We should study them to see what other secrets they hold before they are gone.

***

THE EXTINCTION EVENT

This day may have seemed like any other day in North America during the ice age. Birds sang and lions stalked their prey, but this was the fateful day when the Earth would come across the path of a comet.

The incandescent trail of fire from the passage of the comet through Earth's atmosphere in combination with the fireball from the powerful impact produced a pulse of radiation that set fire to forests and prairies within many kilometers from ground zero. The shock of the impact shattered the glacier and sent ice chunks into the sky powered by the force of the extraterrestrial impact and the steam clouds generated from the glacier ice. All living things within 100 kilometers from the impact died instantly. They were either burned by the heat blast or killed by the shock wave of the impact.

On the East Coast, 1000 kilometers from the impact zone, the blinding flash on the horizon was followed by a sky that darkened ominously as it filled with the giant ice boulders ejected by the impact. Three minutes after the flash, the dark sky advanced relentlessly, and the ground shook as the first seismic waves from the extraterrestrial impact site arrived traveling at 5 km/sec.

By this time, all animals and humans were aware that something terrible was happening. The sky continued to darken, and then the sky filled with bright streaks as the ice boulders in suborbital flights re-entered the atmosphere at speeds of 3 to 4 km/sec.

The ice and debris ejected by the comet formed an inverted cone that swept outward from the impact point. The giant ice boulders started falling in an ever-expanding circle accompanied by sonic booms. The thumping of the ice impacts sent shock waves through the ground that traveled at 5 to 8 km/sec. The humans and the animals could feel the ground tremors of far-away impacts before the ice boulders arrived.

The shaking ground started to liquefy, trapping everyone. The ground had turned to quicksand, making it impossible to walk or run. In a few seconds, everyone would be dead. When the curtain of ejected ice expanded to engulf the East Coast, the ground started shaking ever more violently as the impacts of the boulders got closer. At the peak of intensity, a hail of glacier ice chunks, many as big as a baseball stadium, left steam trails in the sky as they reentered the atmosphere at supersonic speeds and crashed into the liquefied ground accompanied by the thunder of sonic booms. The impacts created oblique muddy conical craters that swallowed whole villages and buried all the vegetation. The tall pines that disappeared in the mud were like minuscule toothpicks compared to the enormous conical craters with diameters of one to two kilometers. The vibration of the ground quickly reduced the depth of the conical craters and turned them into shallow depressions.

Ten minutes after the comet impact, the comet bombardment was over, but it would be about one hour before the rumble of distant sonic booms would cease. The comet itself had not killed the megafauna.

The saturation bombardment by the ice boulders that were ejected when the comet struck the Laurentide sheet caused the extinction event.

The ice boulders hit the fauna directly and destroyed their habitat. The effects might not have been as deadly if the comet had struck hard ground because rock is three times denser than ice and the rock ejecta would have been less voluminous and traveled a shorter distance.

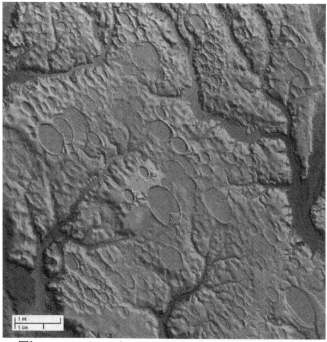

There was no place to hide from the barrage of ice boulders near Tatum, South Carolina

The fires started by the fireball from the impact

73

were probably soon extinguished by the ice bombardment and a fog of ice particles close to the ground. The landscape of the Eastern Seaboard had been transformed into a barren wasteland full of huge, shallow mud holes. All the vegetation had been buried. There was nothing green. All the animals and people were also buried. If there were any surviving stragglers that miraculously escaped the barrage of ice chunks, they would soon be dead. There was nothing to eat. There was no shelter. The rivers and streams in the East Coast had disappeared under a layer of mud that also killed the fish.

The Carolina Bays have remained as evidence of the glacier ice impacts on the soft sandy soil of the East Coast. No such evidence remains of the ice chunks that must have fallen on harder ground, but the ice impacts in the central and Midwestern states were equally merciless. When the colossal chunks of glacier ice hit the hard terrain, they shattered and sent out ice fragments at high speed. Any creature or vegetation in the path of the fast-moving ice shards was destroyed. When the ice finally came to rest, the ejecta blanket had covered one-half of the contiguous United States with a thick layer of crushed ice. The horrific hailstorm left a devastated, lifeless landscape. The layer of ice would take many months to melt in the cold climate of the ice age. The buried vegetation would freeze or remain dormant under the ice. Grazing animals that had survived the glacier ice bombardment had no access to their normal food sources and would soon starve. Predators that were still alive would also soon die without their herbivorous prey.

The days were now dark. The ice crystals from the water and mud that had been explosively ejected above the atmosphere by the comet impact now orbited the Earth and dimmed the light of the Sun. The ice crystals would remain in low Earth orbit for centuries. The ejecta blanket had created an ice cover over a large portion of the United States that increased the albedo of the Earth and reflected a significant portion of the dimmer light from the Sun back into space. The combined effect of the increased ice cover and the orbiting ice crystals would make the land cold and inhospitable for many years.

Eventually, North America would be repopulated by new land animals and new humans, but the megafauna and the ingenious Clovis people that had crafted such fine stone projectiles were gone forever. Not all the species vanished immediately. The megafauna that survived away from the impact zone became more vulnerable to extinction from the sudden change in environment, from the lack of pasture or prey, and from hunting by the expanding human population. The events that led to the disappearance of the megafauna and the fate of the Clovis people could not have been reconstructed without the clues provided by the Carolina Bays.

***

EPILOGUE

The Glacier Ice Impact Hypothesis explains the mechanism by which the Carolina Bays could have formed from ice ejected by an extraterrestrial impact on the Laurentide ice sheet. The elliptical shapes with raised rims resulted from oblique impacts of glacier ice boulders that made conical depressions on sandy saturated soil. In essence, the Carolina Bays are craters made by secondary impacts that were modified by viscous relaxation. The interpretation of the bays as mathematical conic sections provides guidance for future geological exploration.

This book presents a case for "resurrecting" the Younger Dryas Impact Hypothesis using a solid scientific foundation that incorporates data gleaned from the structure and distribution of the Carolina Bays.

The first order of business would be to prove once and for all that the Carolina Bays were formed by impacts. The raised rims of the bays should be examined for inverted stratigraphy that is characteristic of impacts. This may be difficult or impossible because the roots of the vegetation growing on the rims may have penetrated beyond the height of the rims. Similarly, the search for stones that could have been imbedded in the chunks of ejected glacier ice may also be difficult. Many ice boulders would not have contained stones, and the ice boulders that carried stones could have penetrated about 300 meters below the surface for bays with a major axis of one kilometer.

In tandem with geological explorations, it would

be useful to map all the Carolina Bays, to confirm and expand the work done by Davias. A database containing all the bays with their sizes, ellipticities and orientations could be used to get a more precise idea of the location of the extraterrestrial impact and calculate the amount of ice ejected. Assuming a circular distribution of ejecta, the area covered with Carolina Bays represents only 16 to 20 percent of the total ice ejected by the impact.

The time has come to consider using computer simulations to determine the conditions that would have been required for the formation of the Carolina Bays as secondary impacts. Melosh and Pierazzo (1997) have noted that phase transitions can cause long delays in the formation of vapor plumes. This is particularly important for an impact on ice because the process involves two phase changes as the ice first forms water, which is then converted to steam. Such simulations may be able to explain the low angles and large area covered by the ice boulders that created the bays. It has been suggested that expanding vapor plumes may accelerate impact crater ejecta to high speed (Melosh 1989 p. 68). The simulations may also be able to provide more accurate estimates of the amount of ice ejected and could also be applicable to impacts on icy planets.

We may never know some things for sure. With so much ice being launched ballistically above the atmosphere by the extraterrestrial impact, it is very likely that large amounts of liquid water also followed the ice chunks into the vacuum of space. There, the water would have turned into ice crystals that orbited the Earth and blocked the light of the Sun for many years. This reasonable deduction may be hard to prove because we only have the Younger Dryas cooling event as evidence.

For the skeptics who still doubt that oblique impacts make conical cavities that transform into elliptical bays with raised rims, I highly recommend filling a tub with equal parts of pottery clay and sand. Add water and mix until the mixture has the consistency of mortar. Then, start firing ice cubes at the viscous mixture with a slingshot. Shake the tub to speed up viscous relaxation. It is educational and a lot of fun!

***

REFERENCES

Agenbroad, Larry D., The Hudson-Meng Site: An Alberta Bison Kill in the Nebraska High Plains, June 1989, ISBN-13: 978-0962475009

Broecker, Wallace S. (2006). "Was the Younger Dryas Triggered by a Flood?". *Science* 312 (5777): 1146–1148. doi:10.1126/science.1123253, PMID 16728622

Brooks, M. J., B. E, Taylor, and A. H. Ivester, 2010. Carolina bays: time capsules of culture and climate change. *Southeastern Archaeology*. vol. 29, pp. 146–163.

Burney, D. A.; Flannery, T. F. (July 2005). "Fifty millennia of catastrophic extinctions after human contact", *Trends in Ecology and Evolution* (Elsevier) 20 (7): 395–401. doi:10.1016/j.tree.2005.04.022, PMID 16701402

Clark, Peter U. and Alan C. Mix, 2002, Ice sheets and sea level of the Last Glacial Maximum. *Quaternary Science Reviews* 21(1-3), pp. 1-7.

Cooke, C.W., 1954, Carolina Bays and the shape of eddies. USGS Prof. Paper No. 254: 195-206.

Davias, M., Carolina Bay LiDAR Imagery Viewer http://cintos.org/SaginawManifold/GoogleEarth/LiDAR_Viewer/index.html

Davias, M.; Gilbride, J.L.; 2010, Correlating an Impact Structure with the Carolina Bays, *GSA Denver Annual Meeting* (31 October - 3 November 2010), Paper No. 116-13

Dillehay, Tom D. (11 May 1999). "The Late Pleistocene Cultures of South America".

Evolutionary Anthropology, 7 (6): 206–216. DOI: 10.1002/(SICI)1520-6505(1999)7:6<206::AID-EVAN5>3.0.CO;2-G

Dulik, Matthew C., et al., "Mitochondrial DNA and Y chromosome variation provides evidence for a recent common ancestry between Native Americans and Indigenous Altaians." *American Journal of Human Genetics* 90 (2012): 229-46.

Eimers J.L.; Terziotti, S.; Giorgino, M.; (2001), Estimated Depth to Water, North Carolina, Open File Report 01-487
http://nc.water.usgs.gov/reports/ofr01487/

Eyton, J.R; Judith I. Parkhurst A Re-Evaluation Of The Extraterrestrial Origin Of The Carolina Bays, Geography Graduate Student Association, University of Illinois, Urbana Champaign, 1975.
http://abob.libs.uga.edu/bobk/cbayint.html

Firestone, R.B., et al., 2007. Evidence for an extraterrestrial impact 12,900 years ago that contributed to the megafaunal extinctions and the Younger Dryas cooling. *PNAS* 104, 16016–16021.
http://www.pnas.org/content/104/41/16016.full

Firestone, R.B., The Case for the Younger Dryas Extraterrestrial Impact Event: Mammoth, Megafauna, and Clovis Extinction, 12,900 Years Ago. *Journal of Cosmology*, 2009, Vol 2, pages 256-285.
http://journalofcosmology.com/Extinction105.html

French B.M. (1998), Traces of Catastrophe: A Handbook of Shock-Metamorphic Effects in Terrestrial Meteorite Impact Structures. LPI Contribution No. 954, Lunar and Planetary

Institute, Houston.

Garvin, J. B., et al., "Linne: Simple Lunar Mare crater geometry from LRO observations", 42nd Lunar and Planetary Science Conference (2011) http://www.lpi.usra.edu/meetings/lpsc2011/pdf/2063.pdf

Grayson, Donald K.; Meltzer, David J., Clovis Hunting and Large Mammal Extinction: A Critical Review of the Evidence, *Journal of World Prehistory.* Dec 2002, Vol. 16 Issue 4, p313-359.

Greeley, R.; Fink, J.; Snyder, D. B.; Gault, D. E.; Guest, J. E.; Schultz, P. H., Impact cratering in viscous targets - Laboratory experiments, Lunar and Planetary Science Conference, 11th, Houston, TX, March 17-21, 1980, Proceedings. Volume 3. (A82-22351 09-91) New York, Pergamon Press, 1980, p. 2075-2097.

Halligan, J., et al. Pre-Clovis occupation 14,550 years ago at the Page-Ladson site, Florida, and the peopling of the Americas. *Science Advances.* Vol. 2, May 13, 2016, p. e1600375. doi:10.1126/sciadv.1600375.

Johnson, D, The Origin of the Carolina Bays, Columbia University Press, 1942.

LeCompte, Malcolm A., et al., Independent evaluation of conflicting microspherule results from different investigations of the Younger Dryas impact hypothesis, *PNAS* 2012 : 1208603109v1-10.
http://www.pnas.org/content/early/2012/09/12/1208603109.full.pdf+html

May, James H., and Andrew G. Warne, Hydrogeologic and geochemical factors required for the development of Carolina Bays along the Atlantic and Gulf of Mexico, coastal plain, USA, *Environmental & Engineering Geoscience*, August 1999 v. 5 no. 3 p. 261-270

Melosh, H.J., 1989, "Impact Cratering: A Geologic Process", Oxford University Press.

Melosh, H. J.; Pierazzo, E., 1997, Impact vapor plume expansion with realistic geometry and equation of state, Conference Paper, 28th Annual Lunar and Planetary Science Conference, p. 935.

Melosh, H.J; Beyer, R.A., 1999, Computing Projectile Size from Crater Diameter.
http://www.lpl.arizona.edu/tekton/crater_p.html

Melton, F. A., and Schriever, W. 1933. "The Carolina 'Bays' - Are They Meteorite Scars?" *Journal of Geology*, Vol. 41, pp. 52-66.

Meltzer, David J.; Vance T. Holliday, Would North American Paleoindians have Noticed Younger Dryas Age Climate Changes? *J World Prehist* (2010) 23:1–41, DOI 10.1007/s10963-009-9032-4

Petaev, Michail I.; Shichun Huang, Stein B. Jacobsen, Alan Zindler, Large Pt anomaly in the Greenland ice core points to a cataclysm at the onset of Younger Dryas, *PNAS* July 22, 2013, doi: 10.1073/pnas.1303924110

Pinter, Nicholas; Andrew C. Scott; Tyrone L. Daulton; Andrew Podoll; Christian Koeberl; R. Scott Anderson; Scott E. Ishman; The Younger Dryas impact hypothesis: A requiem, *Earth-Science Reviews*, Volume **106**, Issues 3–4, June 2011, Pages 247–264.

Prescott, G. W.; Williams, D. R.; Balmford, A.; Green, R. E.; Manica, A., Quantitative global analysis of the role of climate and people in explaining late Quaternary megafaunal extinctions, *PNAS*, 2012, vol. 109, issue 12, pp. 4527-4531

Prouty, W. F., 1952. Carolina Bays and their Origin, *Bulletin, Geological Society of America*, vol. **63**, pp. 167-224.

Schulson, Erland M.; The Structure and Mechanical Behavior of Ice, *Journal of The Minerals, Metals & Materials Society JOM*, 51 (2) (1999), pp. 21-27. http://www.tms.org/pubs/journals/JOM/9902/Schulson-9902.html

Schweickart, Russel (Astronaut) talking to Peter Warshall about waste disposal in space. *"Watershed Issue" (Winter 76-77) of The CQ.*
http://settlement.arc.nasa.gov/CoEvolutionBook /SPACE.HTML

Shoemaker, E. M., 1960, Penetration mechanics of high velocity meteorites, illustrated by Meteor Crater, Arizona International Geological Congress, 21st, Copenhagen, pt. 8, p. 418434.

Stanford, Dennis J.; Bruce A. Bradley, Across Atlantic Ice: The Origin of America's Clovis Culture, 2012

Stickle, A.M.; Schultz, P.H., Subsurface damage from oblique impacts into low-impedance layers, *J. Geophys. Res., 117,* E07006,
doi: 10.1029/2011JE004043, 2012.

Wittke, James H; et al., (2013-05-20). "Evidence for deposition of 10 million tonnes of impact spherules across four continents 12,800 y ago" (PDF). *PNAS*. E2088–E2097,
doi: 10.1073/pnas.1301760110

Zamora, Antonio, Interpreting Carolina Bays as Glacier Ice impacts, June 28, 2013, http://www.scientificpsychic.com/etc/carolina-bays/carolina-bays.html

Zanner, W; Kuzila, M.S., 2001, Nebraska's Carolina Bays, (GSA Annual Meeting, 2001)

ABOUT THE AUTHOR

Antonio Zamora has a multidisciplinary background in chemistry, computer science and computational linguistics. He studied chemistry at the University of Texas, and Computer and Information Science at Ohio State University. During his service in the U.S. Army, Mr. Zamora studied medical technology and worked in hematology at the Brooke Army Medical Center. Mr. Zamora worked for many years as an editor and researcher at Chemical Abstracts Service developing chemical information applications. He also worked as a senior programmer at IBM on spelling checkers and novel multilingual information retrieval tools. He was the author of 13 patents. After his retirement from IBM, Mr. Zamora established Zamora Consulting, LLC and worked as a consultant for the American Chemical Society, the National Library of Medicine, and the Department of Energy to support semantic enhancements for search engines. Mr. Zamora has been interested in astronomy since childhood when his father helped him build a refracting telescope. During retirement, Mr. Zamora has completed massive open online courses in astronomy, geology and paleobiology. He regularly attends the seminars of the Department of Terrestrial Magnetism at the Carnegie Institution of Washington.

Printed in Great Britain
by Amazon